Pond and Stream

More Arthur Ransome's

NATURE BOOKS FOR CHILDREN

available from Woods Creek Press

- The Child's Book of the Seasons
- A Child's Book of the Garden

Pond and Stream

BY

ARTHUR RANSOME

With Illustrations Adapted
from Frances Craine

WOODS CREEK
PRESS

Portland, Oregon

First edition published by Anthony Treherne & Co., London, 1906

Published in 2026 by Woods Creek Press, an imprint of Ad Fontes Press
www.WoodsCreekPress.com

Printed in the United States of America or
manufactured in Canada, the United Kingdom, South Africa, India, or Australia

35 34 33 32 31 30 29 28 27 26 2 3 4 5
ISBN-13: 978-1-946138-04-0 (paperback)
ISBN-13: 978-1-946138-05-7 (ebook)
ISBN-13: 978-1-946138-06-4 (audio book)

Publisher's Cataloging-in-Publication Data

Names: Ransome, Arthur M., 1884–1967, author. | Craine, Frances, illustrator. | Woods Creek Press, compiler.
Title: Pond and Stream / by Arthur Ransome.
Description: Portland, OR : Woods Creek Press, [2026] | Introduces young readers to freshwater natural history through observations of ponds, streams, ditches, lakes, and rivers. The book describes the common plants and animals of these habitats and encourages children to study nature directly, even through the creation of a simple aquarium for continued observation.
Identifiers: LCCN 2026915326 | ISBN 978-1-946138-04-0 (paper) | ISBN 978-1-946138-05-7 (ebook) | ISBN-13: 978-1-946138-05-7 (audio book)
Subjects: LCSH: Natural history (General) — Juvenile works | BISAC: JUVENILE NONFICTION / Science & Nature / Earth Sciences / Water (Oceans, Lakes, etc.) | JUVENILE NONFICTION / Science & Nature / Environmental Science & Ecosystems
Classification: LCC QH48 .R36 2026 (print) | LCC QH48 (ebook) | DCC 578.768—dc23
LC record available at https://lccn.loc.gov/2026915326

♾ This paper meets the requirements of ANSI/NISO Z39.48-1992 (Permenence of Paper).

For Molly.

CONTENTS.

I

ABOUT THE BOOK

THIS is a book about the things that are jolly and wet: streams, and ponds, and ditches, and all the things that swim and wriggle in them. I wonder if you like them as much as they are liked by the Imp and the Elf? You know all about the Imp and the Elf, do you not? Those two small jolly children, who live in a little grey house in a green garden, and know the country and all the things in it, almost as well as they know each other? The Imp and the Elf love everything that is wet. They paddle in the streams, and build dams, and make waterfalls, and harbours, and sail boats, and do all the other things that every sensible person wants to do. And they love all the fishy people who live in the water, and the beasts that crawl in the mud, and the birds that hop from stone to stone in the stream.

At home they keep a big glass tank on one of the bookcases in the study. And that is the aquarium. It is a kind of indoor watery home for the people whom they meet when they mess about in the duck-pond, or the becks that trickle down the valley. You know what a beck is? The Imp and the Elf are north country children, and they would not understand you if you called the beck a stream.

I will tell you about some of the guests who come to stay with us, and live in the watery tank. But they must be talked about at the end of the book. For just now I want to tell you about the ponds and streams from which they come, and the things that have happened to us there, and all the other things that you will want to know, and the things the Imp and the Elf, who are sitting side by side in my big chair, say must be told to you.

II

THE DUCK POND

The Duck Pond is far away at the other side of the village. We walk a mile down over the fields, till we come to the village, and then we go through a little cluster of grey houses, past the tavern with the the picture of the prancing Blue Unicorn hanging out over the door, past the little grey church with the red tiled roof, past the farmyard by the smith's, where there is always a large sized piebald pig grunting in the yard, and out again into the fields. And then, on the left hand side of the road, we come to three stacks, a horse trough, and a piece of commonland.

The common is rough and untidy, with clumps of gorse and thistles and nettles. There is usually a spotty pony chewing the grass, and a goat with naughty looking horns and a grey beard. A tiny donkey with an enormous voice is tethered to a stake in the ground. There is a crowd of geese, who throw out their long necks in vicious curves, and hiss at strangers and sometimes frighten them. They do not hiss at us. Perhaps they know that we would not be very frightened if they did. The Elf likes this last part of the walk, because she loves to imagine she is a goosegirl in a fairy tale, who drives geese, until she meets a noble Prince, who finds out that really she is a Princess all the time. Some days the Imp is quite ready to pretend to be the Prince, and act the whole story. But other days he is in a precious hurry to get to the pond, and the poor Elf has to be a goose-girl without a Prince, and that is a poor business. She soon tires of it, and runs after us across the common.

Long before we reach the pond, we hear the quaack, quaack of the ducks, and see them waddling along with their bodies very near the ground by the muddy edges of the water, flopping hurriedly first on one leg and then on the other. When we get near them we can see that as they lift their feet they turn their toes in in a manner that shows they have not been at all properly brought up. But then without warning they throw themselves forward along the water, and swim, looking, suddenly, quite graceful. Everything looks quite graceful in its proper place, and almost everything looks silly when it is anywhere else. Even swans, who are the most beautiful of all birds in the water, look as ungainly as can be when they walk along the ground. And if you put a fish, who swims beautifully in the pool, out on the dry land, he just flops and dies, and that is not a pretty sight at all.

The duck pond is very big and round. One bank of it is covered with dark trees that overhang and make green pictures of themselves in the water when the wind is still. And partly under the trees, and partly at one side of them, the bank is high and over-hanging and sandy, and in the sand there are little holes where the sand-martins have their nests. The sandmartins are rather like swallows, only instead of building clay nests under the roof edges of a house, they bore holes with their beaks in banks of earth, and make their nests inside them. A very, very long time ago, we used to do just like them, burrowing into the ground, making a passage with a cave at the end of it, and living there under the earth. There are some of these old homes of ours still left in some parts of the country. The Imp and the Elf are fond of the sandmartins, because they are always in a hurry like themselves. It is fine to see them fly swift and low over the pond, and flutter at the mouth of the hole, and then vanish into it, like mice into a crevice in the wall.

But the birds who matter most of the Duck Pond People, are, of course, the Ducks. There are brown ducks, and white ducks, and speckly ducks, and broods of golden ducklings, that the Elf is fond of watching. The little ducklings waddle about just like their mothers, opening and shutting their dirty yellow flat bills that are always far too large for their bodies. They

look like bundles of grey fluff, with crooked legs and waggly necks.

Often we lie flat on the green grass by the side of the pond, when the sun is high and hot, and white clouds and a blue sky are reflected in the water of the pond. We lie lazily and watch the ducks swimming about, looking for their food. We see them plunge in from the flat shelving mud, and swim out like a mottled fleet of boats. They move their heads to this side and to that, and suddenly plunge them down into the water, into the rotting leaves and mud that lie at the bottom of the pond. And then, as they swing their head up again, we see that something is going down inside. And sometimes when the thing is big, a young and lively frog, or a wriggling worm, we see it hanging out of the duck's bill, waiting to be flung about, and gulped at, until, at last, it goes politely down.

Ducks swim just like men in canoes, striking out first on one side and then on the other, as if someone inside the duck were driving her along with strokes of a paddle. As we lie on the bank, we can watch the strong neat stroke, and see how the feet turn up to be drawn back ready to strike out again, just as a good oarsman feathers his oars. The really most amazing thing about a duck, though, is to see it when it comes out of the water. You would think it would be wet. But no, it looks quite neat and dry, though it has only just come from swimming and diving its head in the muddy pond. The Imp and the Elf always used to be puzzled at that. And their old nurse had a habit of saying to them:—"Why, to scold you is like pouring water on a duck's back; it does no manner of good." And one day they said to me, "Why does it do no manner of good to pour water on a duck's back?" I did not know then, so we hunted in a wise book and there was the reason, and when we watched a duck a little more carefully than usual we saw the book told the truth. The ducks keep oil in a hidden place in their tails, and oil their feathers with it. That is what they do when they preen themselves. That is how they manage to be always dry. For water will not stay on anything that is oiled, and really, it is just as if the ducks made their feathers into mackintoshes against the wet.

All the time that we are resting after the walk and watching the ducks, we are keeping a look out for other of the Pond People; and pretty soon we are sure to see some of them. The pond is full of floating weed, the tiny round-leaved duckweed, floating in green patches even in the middle of the pond, and the dainty white crowfoot, near the banks. There is more duckweed than anything else, and sometimes it is like a green carpet floating on the water. As we lie on the bank, we see a sudden movement in the duckweed, and something pushes its way up through the weed, like a stick that has been held down at the bottom, and then loosed of a sudden, so that it leaps up to the surface of the water. The whole length of the Imp wriggles with excitement. It may be a frog, or it may be a newt. There never was such a pond as this for frogs, and we can nearly always find a newt, if we want to see one.

Early in the year, about March, when we come over the common to the pond, the

Imp carries an empty jampot, with a piece of string fastened round the rim of it, and looped over so as to make a convenient handle. The Elf carries a little net, made of a loop of strong wire, with the ends of it forced into a hollow bamboo, and a circle of coarse white muslin stitched to the metal ring. As soon as we are well on the common, the Imp runs on ahead, and long before we catch him up we hear him shouting by the edge of the pond, "Here it is. Here! Here!" And we find him pointing eagerly to a big mass of pale brownish jelly lying in the water. Big frogs lie about in the shallows, and flop off into the deeper parts of the pond, as soon as our shadows are thrown across the water. It is at this time of the year that the frogs do their croaking. As the Elf says, "they are just like the birds, and sing when they have their little ones by them." For that great mass of jelly is made up, though you would not think it, of hundreds of little black eggs, each in a jelly coat, and each with a chance of growing up into a healthy young froglet.

When we have poked the net under the jelly, and after a little struggling scooped some of it out on the bank, we can see the black dots that are eggs quite plainly. The stuff is so slippery and hard to hold that we can see that even the birds and water things must find it difficult to manage. We rather think that the jelly helps a little in keeping the tiny black eggs from being gobbled before they have had time to grow up. But in spite of its slippery sloppiness, we get a little of it inside the jampot, and when we have dipped the jampot in the pond to give the eggs some water, and dropped in a wisp of weed, that loses its wispiness as soon as it can float again, we set off on our way home, planning all sorts of things for little frogs, and making frog tales. Frog tales, the Elf says, are best in summer, "they make you feel so cool." But they are not at all bad in the spring when the Imp holds the jampot up so that we can all see in, and wonder which black spot holds the young frog prince, and which the frog esquire.

If we liked, of course, we could come day after day to the pond, and watch the eggs change and grow in the water. We sometimes do this; but it is so much easier to watch them at home, that we take some of the jelly away in the jampot every year, and put it into a big bell jar set upside down, with sand in the bottom of it, and plenty of water and green weed.

After a day or two the little black spots in the jelly become fish-shaped, and give

little wriggles from time to time, and at last come out and away from the jelly, small wrigglers, that swim about, and fasten under the weed in waggling rows.

The wriggler has a great deal to do yet before turning into a frog. The tail part of him becomes clearer, like a black thread with a fine web at either side of it, and the head of him becomes fatter and rounded, like a black pea, and we can see feathery things hanging out from behind it, which are called gills. Until it grows lungs of its own, like any respectable frog, the wriggling, black-headed creature breathes with these. The tail grows bigger and bigger from day to day, and flaps like anything, driving the little black tadpole (for that is what we call it) through the water in the bell jar, as if it were a little boat, swimming under water, with a busy paddle behind.

One day, about this time, when we are looking at the tadpoles in the jar, where it stands on the long bookcase in the study, the Imp says, "I say, Ogre, isn't it time we saw the blood moving?" And then I bring a little microscope, all bright gilt, out of its case in the cupboard. We catch one of the little tadpoles, and lay it on a slip of glass, and look at it down the long tube of the microscope. The tail of it looks huge instead of tiny, and all over it, inside it, we can see little pale blobs running to and fro; and those are the tadpole's blood. The blobs look like wee fishes swimming in narrow canals all over the tadpole's tail. When the Elf has looked as well as the Imp, we let it slip back from the slip of glass into the bowl, and see it flap away, as merrily as before.

The tadpole grows fast now, and soon two little hindlegs sprout out, and the forelegs follow them, and the little creature looks like a frog with a tail, and a very big tail at that. And then the tail begins to shrink, and every day the tadpole is more like a frog, and more like a frog, until, at last, the tail goes altogether, and there in the bell jar is a baby froglet, who is quite ready to crawl out of the water on a floating piece of cork, and begin life as a land and water gentleman instead of a mere fish.

That is the way the frog young ones grow up. Their mother does not bother about them at all. They have to do everything for themselves. And they do it very energetically. So that as soon as they begin to turn into frogs, we take them back to the pond and let them go; for if we kept them we should soon have them hopping all over the house. A house is no place for a little wet frog. He wants a pond or a muddy brook, and plenty of duckweed to hide under.

The duckweed in the pond is stirred by other things besides frogs, as I have told you already. The Elf and the Imp would be very angry with me, if I did not tell you all about the newts. For they are the most exciting of all the watery things that are not simple fishes. They are like water lizards, or like tiny water dragons, with four legs and a waving tail.

The Imp has a very particular admiration for the he-newts, and a fairy story to explain how it is that they dress in more gorgeous colours than their wives. Here is the story: Once upon a time there were two brown newts who lived in a pond. One was a he, and the other was a she, and neither of them knew which was which, or who ought to obey orders. So they swam about, and presently poked their noses up through the water-weed, and explained their difficulty to a gay old Kingfisher, who was sitting in his rainbow cloak on a bough that hung over the water. They both asked the question at once. Only one of them asked about a dozen times, and went on asking, and the other asked just once very angrily, and then said nothing more. So the Kingfisher, who was clever, knew which was which. "Why, you are the he," said the Kingfisher to the angry one, and he took a brilliant feather from his breast and gently stroked the newt from his head to his tail. And then a queer thing happened. A fiery crest appeared all along its back, and its body became emerald and spotted gold; and the little she-newt clapped her hands to see her handsome husband, and now she always does exactly what he tells her. That is all.

Well, you know, in a way that story is true, for the he-newt does really wear those vivid colours and that fiery crest along its back for just one season in the year. He wears them when he is making love to his little brown lady. He makes love gallantly—fighting his rivals like the noble little water dragon that he is.

Newts are not any more easy to keep at home in a bowl than little frogs. They grow up from eggs, just like tadpoles, only instead of losing their tails and changing into frogs they keep their tails to swim with, and remain newts. They are not easy to keep because they are very clever at climbing. Once we did catch two of the brown lady newts, and the Imp fell splosh into the duckweed just as he was reaching out trying to catch a he. He caught the he all right, but then we had to go home best foot first, for the Imp was a lump of muddy wetness. He chattered all the way home all the same, and as soon as he had changed his clothes we all worked together, rigging up the old tadpoles bell jar with a fresh sandy bottom, and good clear water, and a floating island of cork, and a lot of duckweed. Then we emptied the jam pot full of newts into the bowl, and saw them swim gaily about examining their new home. We left them and had our tea. When we came back we looked at them again, and saw a very beautiful thing. Two of the newts had shed their skins. You know how sometimes, walking on the moor, we come across snakeskins, like hollow transparent snakes, when we can be sure that an adder has passed that way and left his old coat behind, and slipped gaily off in brighter clothes. Well, that was exactly what the newts had done. There were the newts swimming about, and there were their old skins, like pale, grey films, floating in the water. We could even see the shapes of their tiny feet and hands in the transparent filminesses.

That was all very well, but next morning, as I was getting ready to come down to breakfast, there was a shriek and clatter on the stairs, and presently the Imp, very red, came bumping in at my door to say that all the newts had vanished from the bowl, and that the housemaid had just met one as she was coming downstairs with a can of water. She had stepped over the newt on the edge of the landing, had seen it, and dropped the water-can over and over down the stairs. Would I please come? The Imp held out his pocket handkerchief with something wriggling in it. "You have got it?" I said. "Yes," said the Imp solemnly, looking back towards the door, "but don't let her know." We ran down over the bedraggled stair carpet and saw the water-can under the coat-stand, and the housemaid crying on a chair, explaining how she had seen an evil thing with four legs to it sitting on the landing. The cook was watching her, with arms akimbo, saying "Ah, me," and "Poor dear," now and again.

We ran on into the study, where we found the Elf feeling under the bookcases and tables, looking everywhere for the lost guests. We never saw the others again. But we took the one the Imp had caught back to the pond, and, as we put it in, made a vow not to keep newts again. They are the most escapable things we have ever

tried to keep. Besides, they look jollier in the pond, and are probably very much happier. As the Elf said, "We should try to think how we should like it if other things collected us." It certainly would not be pleasant to be bottled up in muddy water for the little newts to see. It is far best to leave them alone, and, when we want to see them, to come quietly over the common to the edge of the pond, when we may easily see half-a-dozen water-dragons run out from the soft mud, and swim, with quick, hasty flaps of their tails, and jerky paddlings of their arms and legs, out into the depths of the pool.

When we lean out over the pond and take a handful or a netful of the duckweed, and pull it to pieces on the bank, we find some of the most daintily-shaped snails fastened among the mass of tiny pale stems.

The Imp and the Elf always think that they are like very wee snakes, coiled round on themselves in little flat coils. And really they are just the same shape as those stone snails that were once alive, that grown-up people call ammonites. There is a fairy story about those stone snails that shows how other people beside the Imp and the Elf have thought them like serpents. Up in the north country there was an abbey by the sea, and in the abbey a saint lived called Hilda. And all the countryside was made dangerous for foot passengers by crowds of poisonous snakes. The folk of the country asked the saint to help them; for they could not walk abroad without fear of being bitten. They could not let their children out alone, because of the deadly things. So the saint summoned all the serpents to the abbey and, standing on the abbey steps, she turned them into stone, and as they stiffened they coiled up in flat circles like the little snails we find among the duckweed stems. That is the story, but we know now that these stones that they find are really snails that lived thousands of years ago, and have gradually been changed into stone. The duckweed snails are fine things for keeping water clear and pure, and the Imp and the Elf always have a few of them in their aquarium to prevent the water from growing green and stagnant and unhealthy. But you shall hear all about that in the last chapter of the book.

These round snails are very small, but the duck-pond is full of living things even smaller than they. When we scoop a jampot full of water out of it, and hold it up to the light we can often see wee round emerald balls rolling round the pot. They are so small that we can only see them if we look very carefully, "I should not think there can be any things smaller than those," said the Elf one hot afternoon as she blinked at the jampot in the sunlight. But there are. Why, even inside those wee round rolling balls there are tinier balls rolling and moving round, and these are quite alive, too. And, far, far smaller than these, there are little things in the pond, so little that we really cannot see them at all unless we put them under a microscope. The Duck Pond is like a little world of its own with ducks for giants, and newts for dragons, and all the tiny folk and the little snails for ordinary citizens.

But though so many of the ordinary citizens are so small, it is quite easy to grow rather fond of them. We hardly ever leave the Pond without the Imp or the Elf saying beggingly, "Let's wait till we see just one more water boatman." And then, of course, we wait, and crane our necks over the pond, and take no heed of the quacking of the ducks, or even of the splash of a young frog as he flops into into the water. All our six eyes and our three heads see nothing and think of nothing except the thing we want. And when we see him, what do you think he is? A little dark beetle with a pale ring round him, shaped like a tiny boat, comes up to the surface for air, and waits a moment, and then goes quickly off again, this way and that, rowing himself with two of his legs that are stronger than the others, and stick straight out from his body, like oars from a boat. He is the water boatman, and somehow he is so brisk and jolly that we think he must get more fun out of the pond than any other of the pond citizens. And that is why we always want to see him last, before we walk off over the parched common, and leave the quacking of the ducks to grow fainter and fainter behind us. We like to think of the water boatman cheerily rowing about and diving among the reflections of the trees. He is a fine person to invent stories about during the walk home.

III

STREAM AND DITCH

"You can have more fun with a running stream than with a pond," says the Imp. And that is because the galloping water, that leaps and runs over the pebbles, seems to do things to you all the time, while the water in a pond just stays still and lets you do things to it. A thousand games can be played with moving water; at every game it is as fresh as if no one had played with it before.

The Imp spends some of his jolliest mornings at the side of the beck, that flows down from the moorland, through a little wood not far from the house. Up on the moor it is a tiny stream, except when the big rains come, and then it is a streak of foaming white in the mist on the hill-side. But, when it has left the heather and bracken and drops through the wood, it is like a little swift flowing river, with shelving rocky sides, and boulders in mid stream, and tiny waterfalls and pools and weirs. Below the wood it flows out through the meadowland of the valley, growing wider, and moving slower as it goes. Often as the Imp has been playing with the leaping water, and I have been sitting near by among the shadowy leaves of the trees, hazels and rowans, that swing over its channel, I have heard him sing over to himself the words of a poem which he knows. It is all about a stream.

"I come from haunts of coot and hern,
I make a sudden sally,
and sparkle out among the fern,
To bicker down a valley.

"I chatter over stony ways
In little sharps and trebles,
I bubble into eddying bays,
I babble on the pebbles.

"I wind about, and in and out,
With here a blossom sailing,
And here and there a lusty trout,
And here and there a grayling."

This is not all the song, but only his favourite verses.

The Imp builds stone on stone across the stream, and makes a bridge with a dozen piers, and flat stones laid across. Or he sets a row of big stones in the stream, so that the water gushes between them. Then he piles little stones against them, and fills the joints with moss and earth until he makes a solid dam, so that the stream rises up and up, deeper and deeper, unable to go any farther, until at last it overflows the top of the dam and, rushing down, pulls everything to pieces beneath it. That is really exciting. It is exciting, too, to make little canoes of folded paper, and put a pebble in for ballast, and let them shoot the rapid, as if there were Red Indians inside, skilfully guiding with carved and painted paddles. These are only a few of the running water games.

In a little hollow of the stream, below the waterfalls, where the falling water has churned out a basin for itself, we sometimes see a trout, silvery bellied, and dark of back, with spots along his sides. But the place where we go to look for fish and other water folk is farther down the stream, below the wood and moorland. The beck is tamer down there, and has given up leaping from ledge to ledge, but flows quietly and smoothly, with a rippling song of its own, over a broad pebbly channel between the green meadows.

Footpaths cross the meadows, and where they come to the brook, bridges have been made by simply laying a huge flat slate stone from bank to bank across the water. One of our favourite ways of picnicking is to take our basket of food across the meadows and camp in the long grass close by one of these bridges. For then we get the best of everything. The best of the meadow things, purple orchids, and kingcups, like enormous buttercups twice gilded, and the delicate butterfly orchids, who are rare indeed,

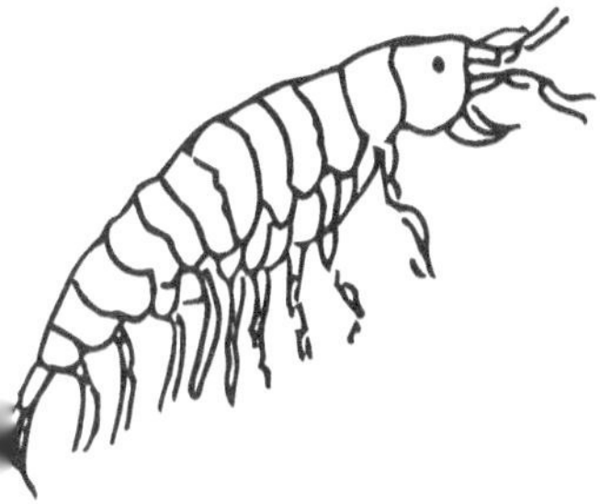

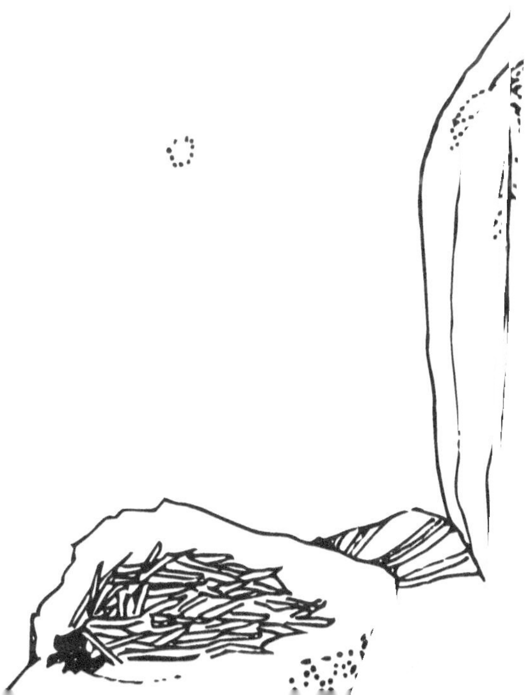

with their pale green spikes, with the white flowers tinted with green, fluttering round them. There is plenty of the little blue forget-me-not growing in clumps close to the water, and ragged robin, with his touzled pink petals close under the meadow hedge. And, best of all, perhaps we see a blue flash, and then another blue flash, and then another, and we know that there is a dragon fly shooting about over the water, and among the water plants, like a small azure comet. Sometimes, when he hovers over a flower, we can see him, but we can never see his wings. They move too fast. And when he is flying about, we can see nothing but the blue glittering flash that shows that he is there.

The best of the water things, too, we get. For lying on the banks of the stream, even while we are eating our sandwiches, we can see the caddises in the muddy shallows, and sometimes a water shrimp, and often a shoal of minnows. And, when the stream is low, the Imp can crawl along, from one side of the bridge to the other, under the big slate, putting his feet and hands on the stones left dry by the water. And that is fun indeed. The Elf and I lie flat on our fronts on the stone bridge, and hang our heads over the edge, and look backwards up the tunnel. And we see the Imp start in at the other end, and come crawling under like a rat in a wet hole. We see his hands and feet clawing about for stones to rest on, and the Elf shouts to him, "There is a stone there—no, there—there, stupid!" and sometimes he finds the stone, and sometimes he does not. We hear him grunt with hotness and excitement. And usually we hear him splash, as one leg or the other slips from its resting-place into the water. And then out he comes, mightily panting, at our end of the bridge. Somehow, with a great pull, he tumbles round on to the bank. And then, because one foot is wet he must take his shoe and stocking off. And if one shoe, why not the other? And if the Imp is allowed to take his shoes and stockings off, why not the Elf? And so, in about three minutes, there are two pairs of stockings and two pair of shoes neatly laid out on the bank, and two small people paddling in the stream, playing for a little, just for the joy of feeling the water stream past their ankles, and then searching about and looking for the little folk of the stream and talking about them, and asking all sorts of questions.

The first and easiest of all the small water folk for people like the Imp and the Elf to find are the caddisworms. Do you know a stonefly when you see one? A long brown-winged dirty-looking fly; you must often have seen one skimming along a brook, and settling on the pebbles that the water has left partly dry. A caddisworm is the thing that is some day going to be a stonefly, just as caterpillars are one day going to be butterflies or moths, and just as Imps and Elves are some day going to be grown-ups. That is all very well. But it does not tell you what a caddisworm is like. This is how the children find one. They paddle to a shallow part of the stream, where it flows under grassy banks, a place where the bottom is a little muddy, instead of being covered with small round pebbles. Then they stand and look into the water, up stream, for the ripples flowing from their ankles make it impossible to see into the water clearly if they look the other way. Then, searching carefully over the bottom, they look for anything small that moves. Presently they see something. It may be a little bundle of tiny sticks, or some pieces of dead grass, or a couple of irregularly shaped twigs, moving crookedly over the sand or mud. And they know that they have found a caddisworm. One or other of them, usually the Imp, dives a hand down into the water and catches it, which is very easy to do, for caddisworms are leisurely people, and do not move much faster than snails. It is lifted out of the water and held out, looking like a little bundle of sticks in the palm of his paw. But while we watch something comes jerkily out of the end of the bundle—a black head and six busy legs, and soon the caddis is crawling along as fast as it can, dragging its house behind it.

For the bundle of sticks is really a log house that the caddis has built for itself. He builds it about his own body all round him, adding stick by stick in the neatest, cleverest manner. He builds with anything he can find, and it is often possible to make him a present of a twig, and see him use it up as a new log in the walls of his house. Nothing comes amiss to him. If the stream he lives in is full of little snails, he is quite ready to cover his home with their shells. Beads, twigs, pieces of grass cut short, flat seeds, scraps of paper, anything you can think of, he will somehow manage to make useful. The odd part of it is that instead of bringing the bricks to his house, or the logs, or whatever you like to call them, he goes in his house to look for each brick, and, when he has finished his building, he carries his home about with him.

As the Imp puts the caddis back into the water he sometimes sees a sudden stirring of the mud, as if someone had poked a pencil in and pulled it quickly out again, bringing a puff of fine sediment up into the water. In the place from which the puff came is a water-shrimp, who is far harder to catch than the caddis, for he is one of the nimblest of the little dodging water-folk. It takes the Imp ten minutes and a lot of splashing before, if he is lucky, he can catch one in the hollow of his hand. Then it lies in a little puddle of water in his palm, whirling itself about, and thrashing into ripples the waters of its prison. It is very like a seaside shrimp, only smaller. It is pale, muddy brown, and looks as if it had been made of tiny napkin rings slipped over each other like a little curly telescope, with active legs and busy feelers.

Sometimes as the children paddle up the stream they see a brown cloud in the water, darting up and up before them in swift swimming jerks. "Minnows!" they shout, and "Minnows, Ogre, look!" and watch the shoal of little fishes flashing through the water just out of reach of them. From moment to moment one of them turns half over in the water, with a flash of silver as he turns. And sometimes, when the Imp and the Elf are not paddling, and we are all three of us lying on the bank, we see the shoal swim slowly past us, and watch the minnows fling themselves right out of the water after the tiny flies that play over the surface of the stream. Then it is as if a clever juggler were hidden under the water and were throwing little curved knives up from the bottom of the stream to twist and sparkle in the air, and then fall plosh, plosh, into widening circles of ripples. Minnow after minnow leaps out of the water, turns and falls, and the ripples of the different splashes cross one another and cut the water into a thousand thousand glittering points of light.

Sometimes we hear a bigger splash than is made by a minnow, and looking up the stream we see something swimming strongly through the water, a double trail of ripples flowing out on either side of him. Just the nose of him is above water, and sometimes he goes under altogether. The thing swims to a flat stone in the middle of the stream that makes a kind of island, and suddenly we see it fling itself up out of the water and sit on its hindlegs on the stone, briskly washing its nose with its fore-paws. "Water-rat," whispers the Imp to the Elf, and we do not move so much as a hair, any

of us. The brown, blunt-nosed rat sits up on the stone and pulls its paws over its head and throws them back again, like the neat-minded gentleman he is, Presently he thinks he hears a noise, an ominous something, and the paws are suddenly still for a moment, and the round head cocked on one side. His head is so blunt and so near his body that one would scarcely think he had a neck at all if he were not able to look this way and that way and this way again, in the smallest part of second. Ah, he sees us. For another instant he stays dead still, wondering perhaps if we have seen him, and then off he shoots again into the water, swimming now on the bottom of the stream and now once more driving his nose along the surface until suddenly he slips under the bank and we cannot see him at all. When we lean over the bank, just where he disappeared, we find a hole, which is the doorway of his home. Here he lives in the moist bank under the over-hanging ferns, close to the water which is as good as land to him. Here he lives and has a merry time to himself, doing nobody any harm, except the waterplants, from whom he takes his dinners.

A little farther down the stream a broad deep ditch crosses the meadows to join it. The ditch is deep, and the water in it moves so slowly that it is almost still. Weeds and grasses grow from the bottom of the stream, and are only just bent over by the current, and the moist edges of the ditch are full of sunken holes, where the cows have thrust their feet into the mud. The whole of the ground by the side of the ditch is rich with flowers, but so swampy that they are difficult to reach, except at a few places. But very often the Imp and the Elf, when their shoes and stockings are once off, make up their minds to despise mud, and wade through the grasses close to the edges of the ditch to look for sticklebacks. And really, when I think of sticklebacks, I agree with the children that it is worth more than muddy ankles to get a look at them. For the sticklebacks are very fine fellows indeed, the little soldiers of the water-people, tiny fishes, who carry spears set upright on their backs, spears that are strong and well pointed, too, as the Imp found when he took hold of a stickle between his finger and thumb.

The sticklebacks are like the newts of the duck-pond in quite a number of ways. Not to look at, of course, for one has legs while the other has fins; but in several of their habits. In the love-making times,

when the he-newts show their gorgeous coats, the stickleback lords put on a brilliant uniform of glittering green and scarlet and gold. Like the he-newts, they battle between themselves, and more than once we have watched a noble skirmish in the deep water under a tussock of grass. We have seen the stickleback lords dash at each other again and again, trying to rip each other up with their sharp spears, and, at the last, we have seen the conqueror sailing proudly away, even more gorgeous than before.

The Imp loves the sticklebacks because they are so bold and jolly and move so quickly and so jerkily that it is hard to follow them. But the Elf loves them for quite another reason. She loves them because they are homely. Most of the water people, like the frogs and newts, take no bother at all about their eggs, but just leave them to themselves without ever caring whether they hatch or no. But the stickleback is as careful as a blackbird, and builds a little nest for the eggs down among the weeds on the bottom of the ditch, and stays there watching and guarding till they hatch out into little stickles. That is why the Elf loves sticklebacks. They do look after their children a little.

Later in the year we see the shoals of little sticklebacks, not so big as pen-nibs, who have left their nests in the ditch, and are swimming away to see the world for themselves. Often we lie on the bank and tell each other stories about them. And all these stories begin: "Once upon a time there was a little stickleback, one of a shoal," and all the stories end: "So the little stickleback drove his enemy away in a fright, and swam back to his nest glowing with colour and pride." For, of course, by the time the story is finished the little stickleback has grown into a big stickleback and has a nest of his own.

Besides the sticklebacks and minnows there are a great many other fishes among the water-folk, but we do not meet them so often, and most of them live in bigger places than the stream or the ditch, or even the duck-pond. Sometimes, though, in the pebbly part of the stream we meet a loach, a little brown speckled fish with a flat head and little suckers like the horns of a snail sticking out all round his mouth. We see him slip along in the water under the shadowy side of a stone. If he does not come out at the other end we know he is resting there, and then if we can make the stone move without muddying the water, we may see him flit from his hiding place, zig-zag among the pebbles, looking for a new stone where he may shelter.

And then, too, when the stream flows nearer to the sea, which is only four miles from our house, you know, we find some other water people who are very pleasant indeed. The sea spreads inland in a broad pale sandy bay, with marshland grown over with sparse reedy grass, and covered with pools of salty water, and channels full of sandy mud. The stream flows out into this bay, and at some times of the year, when we walk up from the bay along its banks, we see stones that look as if they were heads, with a waving mass of black hair flowing from them down the current. When we look closer, we see that the black hair is a mass of tiny eels. Little black wriggling water snakes they look like, though they are nothing of the sort, and we sometimes remind each other of the tale of the Gorgon's Head, with all its snaky crop. Sometimes we have caught a little eel or two, and kept them in a big jar; but they are not such adaptable guests as the tadpoles, and we do not think we make them very comfortable. The Imp loves to watch them, and finds it hard to believe that these are eels, really eels, like the big twisting creatures he sees when he leans over the side of the boat, when we go rowing on the lake. You shall hear about those eels in the next chapter.

But, do you know, I believe our dearest of all the water people, are not really water things at all, but birds? There are two of them, that belong to the stream, and I expect I shall be scolded by the Imp and the Elf for putting them at the end of the chapter. I shall have to explain that I meant it as an honour to them. They are birds; and one of them lives up the stream, where it is a wild little beck, falling from rock to rock in the wood on the moorland side, and the other hops from stone to stone in the shallows of the brook, where it flows more peacefully through the flat green meadows. The one is the dipper, and the other is the water wagtail.

The dipper is a little brown fellow, with a white front and throat, and a jovial little shout of his own. Very often, as we climb up through the wood, with the noise of the thousand tiny waterfalls swishing in our ears, we meet the dipper, perched on a stone by the side of one of the pools, looking as if he were making a careful map inside his head of everything he can see at the bottom of the water. As soon as he notices us, there is a brown flash in the air, and he is up, and over the next fall, and perched on a stone by the pool above. When we have climbed painfully up over the slippery rocks and the soft green earth with the help of hands and knees, and little trees, and clumps of heather, we find him sitting there, as gay and fresh as ever, and perfectly ready to dart up stream again. But sometimes we have been able to watch him, and see him dive into the pool; for he can swim under water as if he were fish and not a bird at all. He can swim round and round the bottom of the pools as easily as he can fly. The Imp thinks him a very fortunate person; for he can do everything. He can swim under water, he can hop about on land, and he can fly in the air. And when you can do all those three things, there is not much else left to want, is there?

The other bird is as dainty and spruce a little fellow as you can imagine. All dark and white he is, looking like a pale and tiny magpie, with a long tail. His tail gave him his name, and I have been told a story about that. Here it is:—Once upon a time there was an old wise man, and he set himself to write a huge book about all the birds that ever are. So he went out with a lot of pens and ink and paper, and lived in a hut at the edge of the meadows, just sheltered by a wood. He told all the birds he knew what he was about, and they told all the others. So that they all came—albatrosses, and sparrows, and thrushes, and penguins, and blackbirds, and guillemots, and seagulls, and flamingoes, and peewits, and ostriches, and kingfishers—and fluttered and chattered in a huge crowd in the meadows by the hut. One by one they perched on a log in front of the old man, and he wrote down what they were like, and what were their names, and all about them. And this all worked very well until he came to the wagtail, when he could not think of a name for it. He put his head on one side and looked at the little mottled bird, and he said, "Well, my life, I do not know what to call you," and the little bird wagged its tail. The old man scratched his head, and said, "Well, you little speckled thing, what am I to call you?" and the little bird wagged its tail. The old man grunted and groaned, and made all the noises we all make when we are stuck over a very simple thing. He could not think of what to write, and he kept dipping his pen in the ink, and scratching his head with the other end of the penholder; and all the time the little bird wagged its tail. Its wagging muddled

the old man worse than before, and he said angrily, "You do nothing but wag your tail, wag your tail, wag—your—tail" and suddenly he found that he had written down wagtail without thinking. And the little bird has been called a wagtail ever since.

"And it does wag its tail all the time," says the Elf. It really does. We see it flit about the shores of the stream, first a little this way, and then a little that, and every time it perches its tail wags up and down, up and down, like a tiny see-saw that has lost its other end.

Sometimes late in the summer we see yellow wagtails by the stream, and they are even prettier than the grey ones, the very daintiest of little fairy birds. But in autumn, both the grey wagtails and the yellow ones fly away over seas like the swallows, and we do not meet them by the stream side till next year.

IV

LAKE AND RIVER

One month in every year the Imp and the Elf and I go to stay in a farmhouse close by the shores of a lake, that is bigger than the biggest pond you have ever seen. Out of the lake flows a river, that is bigger than the biggest stream. But when we go there we always feel very much like we do when we go over the common to the duck pond, or follow the beck from the woodland to the valley. Only, now, instead of lying on the bank at the side of the water, we go in a boat, and row out with the water lapping round us. It is as if we were in an enormous ship of our own, and quite safe, for the boat is so broad in the beam that not even the Imp or the Elf could tumble out if they tried.

Of course we are pirates, and Sir Francis Drakes, and vikings, and other sea rovers, from time to time. I often find that I have been a villainous pirate mate, when, for all I knew, I had been peaceably reading a book in the stern, and we none of us know when we set out in the morning what manner of gay adventures we shall fashion for ourselves upon the water. But, if I were to tell you about all that, I should have no room left in which to write of the water folk, and that would never do.

This is a chapter mainly about the lake things, and they are very like the pond and stream people, only bigger. You remember the little eels we used to find in the stream, clustered like massing black hair below the stones in the running water? Often, when we are floating on the lake, where the bottom is sandy, and not so deep that we cannot see it from the boat, we look over the gunwale and see long brown slimy eels, with silver bellies, twisting along the ground below. Sometimes we drop a worm exactly over an eel, and watch it fall like a curling coral in the water until the eel shoots at it, and gulps it in. Eels are really very like water snakes, but they are fish, with fins just like the trout, and funny little snoutheads, that make us think of pike. Pike are the ugliest and wickedest-looking of all the fish. Big and hungry they are, with evil eyes and long snouts, and mouths set full of teeth that point back down their throats, so that if a little trout or a hand once got inside it has not much chance of escape. The pike lie under the thick weeds round the shores of the lake, and among those rushes that rise out of the water like a forest of green spears. Then when a shoal of big minnows or small perch float by the pike darts out, and there is one perch or one minnow the less in the frightened shoal of little fish. Pike do not like perch very much, because a swallowed perch means a sore pike. For those gay perch with their scarlet fins and golden bodies barred with olive green are not defenceless at all. The fin that runs down their backs is built on firm, sharp spikes that they can lift when they want, and no perch is mild enough to let himself be swallowed without lifting his spines and tearing the throat of the swallower.

As we row on down the lake, watching the reflection of the boat rocking in the ripples, and the reflections of the hills and the trees on the shore of the lake we sometimes hear a long whistling cry, and a curlew swings high over our heads from one side of the valley to the other, like a pendulum-bob without a string, his long curved beak stretched out far before him. And sometimes when the weather is going to be stormy we hear a shrill shriek repeated again and again, and see a white cloud of seagulls lift from the marshy shores and flap away and back and settle again. And more than once we have seen wild duck, and a drake in a gorgeous shimmer of colour fly across the marshland by the head of the river. Half-way down the lake there is a little rocky island, where we have often seen the yellow wagtails, and on a promontory opposite a kingfisher has his nest in a deep hole in the rock. We row the boat close up to the nest, and look at the pile of fishbones outside the hole. The Kingfisher is a fisher as well as a king, and lives on the fish that he catches. It is fine to see him fly across in the sunlight from the rock to island, from the island to the rock, "Just like a rainbow without any rain," as the Elf says, for he is the most gaily coloured of the birds. "Because he is the king," says the Imp. And indeed his kingly robes are very splendid—blue, and green, and red, and white, and orange—as fine a cloak for a monarch as you could wish to see.

There is another fishing bird whom we are always glad to see. He is bigger than I do not know how many Kingfishers put together, and though he is not brightly coloured he is very beautiful. He is a heron, and herons are like the storks of Hans Andersen's fairy stories. He is a grey bird, tall and thin, with a black crest lying back from the top of his head. His legs are long, and he is fond of standing on one leg by the edge of the water, or on a stone at the end of some little promontory, tucking the other leg up in the air, and watching the water with his head on one side, ready at any moment to dive his long beak into the lake and snatch a little fish out of it. When he flies he crooks his long neck back on his shoulders and hangs his legs straight out behind, so that it is impossible not to know him when you see him.

In the little harbour, where our boat is kept, there are often so many minnows that when we look into the water it seems as if the bottom were made of moving tiny fish. People who are going to fish for perch often catch the minnows dozens at a time in nets in their boathouses, when the water is shallow, and the minnows swim up into

the shadows of the boats. And there are caddises there too, if we choose the right places to look for them.

As we walk down through the fields from the farm to the boathouse, the Imp and the Elf leaping for joy in themselves, and the sunshine, and the cool wind, and the blue hills, we plan what we shall do with the day, and where we shall go, and whom of the lake people we shall try to see. And one morning or other, as we leave the farmyard, the Imp cries out, "I say, Ogre, isn't to-day the day for a picnic down the lake?" And the Elf says, "Yes. Say yes, Ogre, do," and in three minutes we are all as happy as pioneers arranging an expedition. After lunch we start, and by that time the sandwiches are cut, and the bun-loaves, and the marmalade, and the tea (hot and corked up in a bottle), and the mugs, and everything else are all packed into two baskets by the jolly old farmer's wife, and we go off together, the Imp and the Elf carrying one basket between them, while I carry the other.

We run the boat out of the boathouse, and when we have settled down, the Elf and the baskets in the stern, and the Imp lying flat on his stomach in the bows, we slip away down the lake rippling the smooth waters, and leaving long wavelets behind us that make the hills and trees dance in their reflections. We glide quietly away down the lake, looking up to the purple heather on the moors, and the dark pine-woods that run right down to the water's edge, and watching the fishermen rowing up and down trailing their lines behind them, or casting again and again over the waters of the little rocky bays that break the margin of the lake. That is one way of being interested in the water people; to want to catch them on a hook at the end of a line, but it is not our way. We think of the water folk as we think of the fairies, as of a strange small people, whom we would like to know.

We row down the lake, lazily and slowly, past rocky bays and sharp-nosed promontories, and low points pinnacled with firs. The hills change as we row. At the head of the lake they are rugged and high, with black crags on them far away, but lower down the lake they are not so rough. There are fewer rocks and more heather, and the hills are gentler and less mountainous, until at last at the foot of the lake they open into a broad flat valley, where the river runs to the sea. A little more than half way down there is an island that we can see, a green dot in the distance, from our farmhouse windows, and here we have our tea.

We run the boat carefully aground in a pebbly inlet at one end of the island. We take the baskets ashore, and camp in the shadow of a little group of pines. There is no need to tell you what a picnic tea is like. You know quite well how jolly it is, and how the bun-loaf tastes better than the finest cake, and the sandwiches disappear as if by magic, and the tea seems to have vanished almost as soon as the cork is pulled from the bottle.

As soon as tea is over we prowl over the rockinesses of the little island, and creep among the hazels and pines and tiny oaks and undergrowth. Do you know trees never look so beautiful as when you get peeps of blue water between their fluttering leaves? When we have picked our way through to the other end, we climb upon a high rock with a flat top to it, and heather growing in its crevices; and here we lie, torpid after our tea, and pretend that we are viking-folk from the north who have forced our way here by land and sea, and are looking for the first time upon a lake that no one knew before us. The Imp tells us a story of how he fought with a red-haired warrior, and how they both fell backwards into the sea, and how he killed the other man dead, and then came home to change his wet clothes, long, long ago in the white north. And the Elf, not to be beaten, has her story, too, how she rode on a dragon one night and saw the lake—this very lake—far away beneath her, like a shining shield with a blue island boss in the middle of it. And how the fiery dragon flapped down so that she could pick a scrap of heather from the island, and how here was the very heather that she picked. And then I tell them stories, too, of the old times, when the great fires were lit on the crests of the hills, as warning signals to people far away. And so the time slips away, and we suddenly find that we are ready to row on again to have just one peep at the river.

All round the low end of the lake there are tall reeds growing and bulrushes, and there is soft marshy ground that make damp islets among the reeds. As we row down we are nearly sure to see one or two big white birds with proud necks swimming slowly along the reeds. Sometimes we have seen them rise into the air with a great whirring of wings and splashing of water, and then sink again on the surface of the lake, beating up a long mane of foam as they fall. These are the swans,

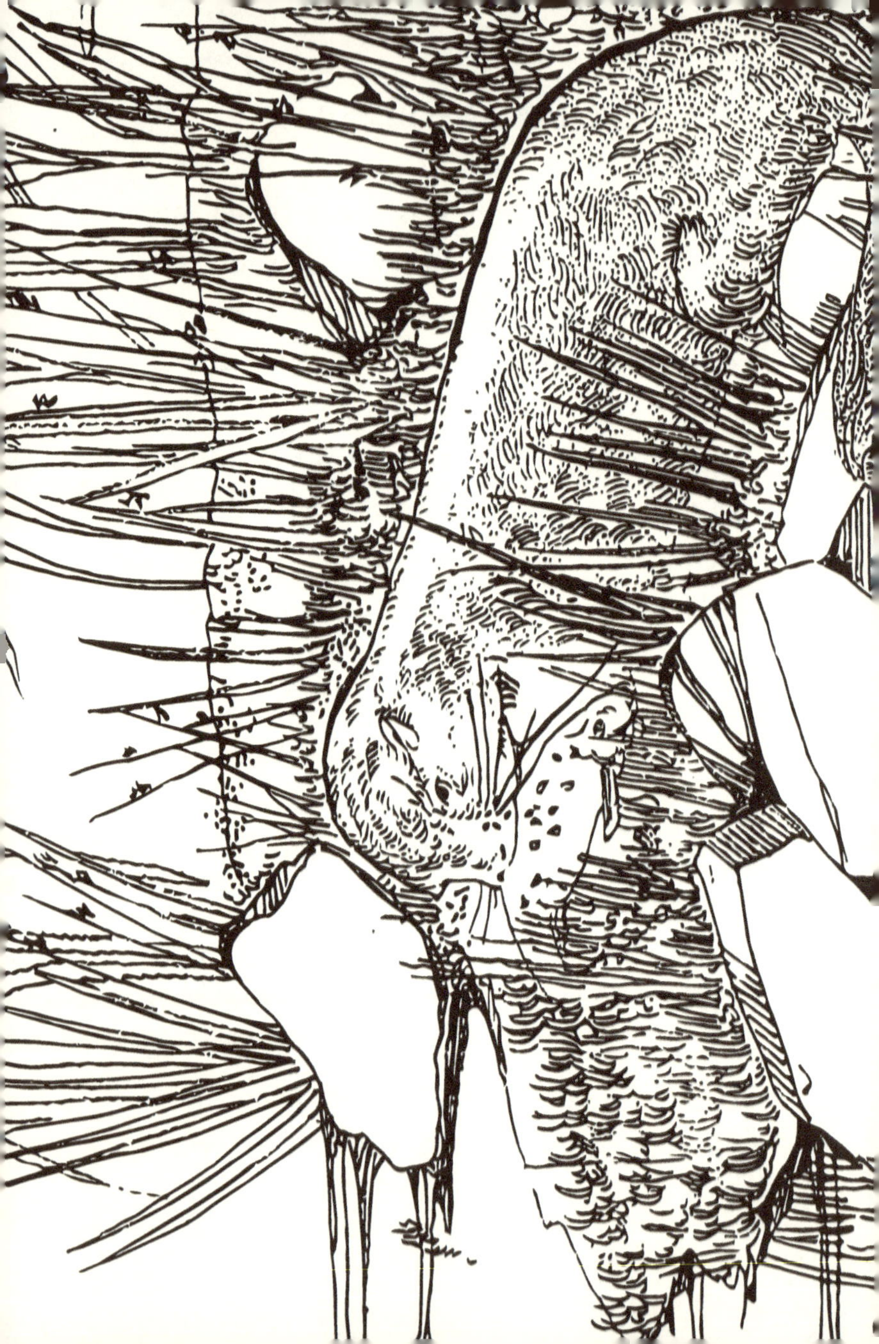

sometimes see the remains of a dead fish half pulled to pieces. We know what that means, "The otter," says the Imp, and we stare about with eyes wider than before, doing our best to imagine in very stir in the bushes or under the banks that we can see his dark body, like a beaver, for he can swim in the water and dive like a fish, and run along the bank as well. But we have never seen him, though we know that he is there. And otters are growing fewer and fewer. Every year men and women with dogs come to hunt them and kill them. Some day there will be no otters left at all.

We wait in the river till the evening, and then set out to row the long way back again. As we row up the river into the lake again we can see the trout rising in big circles of ripples, and hear the peewits screaming on the marshland. It is odd how we seem to notice sounds at evening that we should not at other times. Everything seems so quiet that little noises seem to matter. When we hear the frogs croaking we do not think how loud they are, but only how silent is everything-else. It is evening now, when we row round the promontory at the low end of the lake, and already we are wondering if we shall get home before the owls begin to call. Long ago the Imp and the Elf should have been asleep in bed.

The lake is very still, and the sky is less brilliant than it was. The sun has dropped below the hills, and their outlines are clear against the rose of the sunset. The Imp and the Elf say nothing, but listen for the night noises, and watch the sky working its miracles in colour. This evening is a new dream world for them, and they are wondering whether the water people are awake or asleep. "There never is a time when everything goes to bed, is there?" says the Elf, sleepily, as we lift her out of the boat. And as the two of them go off to bed, very happy and very, very tired, we can hear the long kr-r-r-r-r of the nightjar in the pinewoods up the hills, and below us in the woods at the head of the lake two owls are answering each other.

V

OUR OWN AQUARIUM

It is quite a long time since the Imp and the Elf first started a guest-house for the water people. One day, when the Elf was very small, and I was showing her pictures in a book, and telling her about the sticklebacks, and the minnows, and the loaches, and the caddisworms, and all the rest of them, she sat silent for a long time, and then said suddenly, "I want to ask him," and wriggled down from my knee and went off to find the Imp. Presently they came back together. "We want to have some caddises for our own," they said, and I understood that the Elf had thought it only fair to consult the Imp before asking me about them for herself.

That very day we began to plan the guest-house. At first it was to be no more than a jam pot, with mud in the bottom of it for the caddises. Then we thought that perhaps even a caddis would like a house a little bigger than a jam pot, or even than a big marmalade jar. Even caddises crawl. The next bigger thing to one of the big marmalade jars that they have in the nursery is a basin. And basins are no use at all. They tip over if you lean on their edges to look at anything that is crawling about inside. There was nothing for it but to plan something new. And, if we were to have something made on purpose, if we were to have a really big home for caddises, there was no reason why we should not plan it bigger still and be able to keep minnows in it, or goldfish, or even a smallish eel.

So we spent a splendid afternoon planning the guest-house, and next morning walked over the fields to the village with a lot of scribblings in our hands. The scribbles were to explain what sort of a guest-house we wanted. We walked straight through the village to the glazier's shop. A glazier is a man who comes and mends windows when tennis-balls have gone through them and broken them. This glazier was very nice and kind. He let the Elf and the Imp climb up and sit on his table, while he looked over our scribbles, and then took a big sheet of paper and made a neat drawing himself. He made what he called a plan, and what he called an elevation, and then he drew a real picture of what the guest-house was to be, and put a curly fish with a winking eye swimming about in the middle. This picture he gave to the children, so that they could think about the guest-house while it was being made. He promised that we should have it in a week's time.

It was a fortnight before it came. That is the way of glaziers who are leisurely but very clever. For though the guest-house was so long in coming, it was splendid when it came. It had four sides made of glass, with wooden pillars at the corners, painted green. It was like a house whose windows had spread all over the walls. And it was so big that the Imp could easily stand in it with both feet, a good way apart, too. We filled it with water and it did not leak. There was a tube hidden in the bottom of it with a tap at the side, so that we could let the water out and put fresh water in without having to take out the fish. That

chapter. We wanted plenty of them, because they keep the aquarium healthy, and the water sweet and fresh. As soon as we had plenty of duckweed and plenty of snails, we went on over the fields to the beck. And here we got half-a-dozen caddisworms and a water-shrimp, and some minnows. We let the shrimp go, because he does not live well except in running water. But the others we carried home with us in the jam pots, which we had to pretend into triumphal carriages. For we were bringing home our first guests.

The Imp and the Elf sat on high chairs in the study till bed-time watching the caddises crawl about on the mud, and the minnows flit in and out among the stones. And before they went to bed they said goodnight, very solemnly, to the water people. For it is always best to be polite, even if the water-people do not understand. And, as the Elf said, "Perhaps they do."

Next morning the carrier stopped on his way from the station with a big can that had come by train from London; and in the dark depths of the can we could see golden flashes. For I had written to town for half-a-dozen golden fish to come and stay with the minnows.

And after that the guest-house was always full. From time to time new guests came, and others went away, let loose again in the duck pond or the stream. Always the guests are changing. Someone sent us a little water-tortoise for a present, and we kept him with us for a little while, and then put him in the pond to see life on his own account. We have had little eels from the stream, and sticklebacks (but these are quarrelsome folk), and tadpoles, and loaches, and carp, who are like greenish goldfish, and long-bodied gudgeon, and silvery roach. Every morning, after breakfast, before setting out walking, the children come into the study and feed the guests with worms, and ants' eggs, and crumbled vermicelli.

The guest-house is like a little water world where we can see the smaller water-folk living in their own way. It is a beautiful little world, with its clear water, and green weed, with the little fishes swimming under the roots of the weeds, and darting among the crevices of the stones. And it is a little world that is not very difficult to manage. We have to be careful not to overfeed the guests, and yet we must be sure that they have enough to eat. We have to keep the water clear, changing it every other day, pouring fresh water in at the top and running out the old through the tap at the bottom. It is a little world that anyone can manage who loves the water-folk well enough to take plenty of trouble with them.

❁ ❁ ❁ ❁ ❁

And now, do you know, we have come to the end. There is such a lot to write about the things that are jolly and wet that the Imp and the Elf say I have missed out half the things that ought to be put in, and I know that I have missed out a very great deal more than that. But if you really care for the water-folk you will find out the best of all the things that cannot be written here by going to the stream side, or the pond side, or the side of the lake and making friends with the water-people for yourself.

ABOUT THE AUTHOR

Arthur Ransome was born in Leeds, England to Cyril and Edith Ransome on 18 January 1884. He was the elder brother of Geoffrey and of their two sisters, Cicely and Joyce. Sadly, his father, Cyril, died when Arthur was only thirteen years old.

Arthur was not a brilliant scholar in school, but he loved to read and aspired to become a writer. He met William G. Collingwood (in 1903), the author of *Thorstein of the Mere*, a book Arthur had read as a child. Collingwood and his family more or less adopted Arthur, and, in time, Arthur published his first two collections of essays, *The Souls of Streets* and *The Stone Lady*. Neither collection was especially well received, but Arthur later found success with his later books.

Arthur continued writing after he published his *Nature Books for Children*, including this book, *Pond and Stream*, as well as *The Child's Book of the Seasons* and *The Things in Our Garden*. In due course, Arthur married, wrote about other writers, such as Edgar Allen Poe and Oscar Wilde, was sued for libel, travelled to Russia (where he reported on the Bolshevik Revolution in 1917), divorced, remarried—this time to Leon Trotsky's secretary—travelled to China, among other places, and, after some time, returned to England. There he wrote the books for which he is best known: *Swallows and Amazons*, a series of children's adventures set in the beautiful

Lake District of England. This series has never been out of print and has even been adapted into feature-length films, in 1974 and 2019.

In 1953, Arthur was awarded the honour of Commander of the Order of the British Empire (CBE) for his outstanding contributions to British society. Arthur died at the age of eighty-three on 3 June 1967 in Manchester, England. During his life he wrote many short stories, essays, and two unfinished works—*The River Comes First* and *Coots in the North*—which have since been published.

This book was set by Matthew J. Barron.

The text face is set in 14-point Adobe Caslon Pro, a revival of the typefaces of William Caslon by Carol Twombly.

www.ingramcontent.com/pod-product-compliance
Lightning Source LLC
Chambersburg PA
CBHW031334230926
51848CB00001B/229
* 9 7 8 1 9 4 6 1 3 8 0 4 0 *